Each activity was developed to be completed in approximately 40-50 minutes. Many can be adapted for a shorter time frame if needed. Some may also take longer, particularly when completing a Dig Deeper. Insect identification is not covered in this manual. See the Resources page for insect identification resources.

Table of Contents

Chapter 1 Be an Entomologist	**2**
Activity 1: What Is an Insect?	3
Activity 2: Copy Cat	4
Activity 3: Define It	5
Activity 4: Big Mouth Bugs	7
Activity 5: FACETnating	8
Activity 6: Insect Olympics	9
Chapter 2 Biodiversity	**11**
Activity 7: Pit Stop	12
Activity 8: Buzz-z-z-zing Around	14
Chapter 3 Invasive Species	**16**
Activity 9: Alien Insects	17
Activity 10: Establishing a Toe-Hold	19
Chapter 4 Integrated Pest Management	**21**
Activity 11: Where Are They?	22
Activity 12: Ants and Uncles	24
Chapter 5 Forensic Entomology	**27**
Activity 13: Insect Investigation	28
Activity 14: Chirp, Chirp	29
Activity 15: Sherlock Bug	31
Activity 16: I Eat Insects	33
Resources	**35**
Glossary	**36**

Date Activity Completed

Chapter 1: Be an Entomologist

Insects are fun to watch and study. If you are interested in entomology – the study of insects – you can start by learning insect body parts and how insects see and move. While all insects have the same basic parts, there are a lot of differences in how the parts look and work. This is called diversity.

Activity 1: What Is an Insect?

What makes an insect an insect? All insects have a body that is divided into three parts (head, thorax, and abdomen), a pair of antennae, and three pairs of legs. Some adult insects have one or two pairs of wings. In this activity, you will use the main parts of insects to create a new insect.

Mix & Match

Tool Kit

☐ Insect pictures
☐ Drawing paper
☐ Pencil
☐ Color markers
☐ Clear tape

Find pictures of five different insects in books or magazines.

- Using each picture as a guide, make separate drawings of each of the insect body parts listed below. You will end up with a set of drawings for each of your five pictures.

Antennae	Hind legs
Head	Thorax
Front legs	Abdomen
Middle legs	Wings (if present)

- Color the insect body parts.
- Cut out each insect body part.
- Separate the insect body parts into piles (antennas, heads, thorax, etc.) and then turn them face down.
- Then, draw out one body part from each pile and arrange the parts to make a new insect.
- Tape your new insect so it stays together.
- Label each of the insects you created with the following information.

  Title: My New Insect
  Name:
  Where it lives (habitat):
  What it eats:
  Interesting facts:

Life Skills

ENTOMOLOGY SKILL: Learning insect body parts.
SCIENCE STANDARD: Biological evolution.
SUCCESS INDICATOR: Can name the major insect parts.

Talk It Over

SHARE WHAT HAPPENED: What body parts can you name?

APPLY: How is the insect you made like a real insect? How is it different?

GENERALIZE TO YOUR LIFE: How will what you learned help you with future activities with insects?

Activity 2: Copy Cat

Insect body parts come in different shapes, sizes, and colors that result in a variety of body forms adapted to different environments, but the basic body plan of insects is the same. Insects may be small or large, drab or brightly colored, fast or slow, but they all play a role in nature.

Insect Model
Tool Kit

Common materials found your home. For example:
- ☐ Pencil
- ☐ Rubber bands
- ☐ Pipe cleaners
- ☐ Construction paper
- ☐ Milk jug cap or bottle cap
- ☐ Dryer sheets
- ☐ Dried beans or rice
- ☐ Modeling clay
- ☐ Grass blades
- ☐ Leaves (green or dried)
- ☐ Sticks

- Find a picture of an insect you like or use one of the insects that you made in Activity 1.
- Collect common materials found in and around your home.
- Use the materials that you collected to make an insect model.
- Be sure your model includes these insect parts:

Antennae	Hind legs
Head	Thorax
Front legs	Abdomen
Middle legs	Wings (if present)

- Draw a picture, or attach a photo of the insect you created. Identify each part to your helper.

My Created Insect

Life Skills

ENTOMOLOGY SKILL: Learning insect body parts.
SCIENCE STANDARD: Biological evolution.
SUCCESS INDICATOR: Can name the major insect parts.

Talk It Over

SHARE WHAT HAPPENED: What materials did you use to create your insect and how did you assemble it?

APPLY: How could you change your insect model to be like a real insect?

GENERALIZE TO YOUR LIFE: What careers involve decision-making skills?

Activity 3: Define It

All insects have the same basic body parts. Knowing the basic parts allows you to identify insects and their relatives that are not insects. These body parts come in many different styles, sizes, and colors. The differences have evolved to allow insects to live in different habitats throughout the world.

Insect Body Parts

Draw a line between the body part and its definition.

Hint: Match the body parts and definitions that you know. Then, use the Glossary to match the rest.

Body Part	Definition
Abdomen • (ab-doe-men)	• This is the middle body region between the head and the abdomen. It is divided into three segments, each with a pair of legs. If there are two pairs of wings, they are on the last two segments. If there is only one pair, it is on the middle segment.
Antenna • (an-ten-na)	• Found on the last segment of the abdomen of female grasshoppers, it is used to lay eggs.
Compound Eye •	• There are two of these in front of or between the eyes. They are used to touch, smell, hear, and, in some cases, detect moisture and temperature. The shape varies.
Head •	• Most insects have two of these, made up of many individual lenses.
Leg •	• This helps insects hear. They are found on each side of the first segment of the abdomen in grasshoppers, and on the inside of the front legs in crickets. Not all insects have these.
Ovipositor • (o-vi-poz-i-ter)	• Often the largest region, it usually has nine or 10 segments or rings.
Spiracle • (spi-ra-kal)	• Insects breathe through these small holes on the side of the thorax and abdomen. There are two pairs on the thorax and eight pairs on the abdomen in most insects.
Thorax • (thor-ax)	• An insect has three pairs of these: front, middle, and back. Not all three pairs are alike and sometimes the back pair is used for jumping (as found on a grasshopper).
Tympanum • (tim•pa•num)	• This body part is like an upside-down bowl. The compound eyes and antennae are on the top or sides. Mouth parts are on the underside.

TEAMING WITH INSECTS

Activity 3: Define It

Life Skills

ENTOMOLOGY SKILL: Learning insect body parts.
SCIENCE STANDARD: Biological evolution.
SUCCESS INDICATOR: Can match insect parts to their definitions.

Dig Deeper

Collect three insects and see how many insect body parts you can identify for a friend. Record your data in the graph paper section below, or download a data page from the 4-H website at www.4-H.org/curriculum/.

Talk It Over

SHARE WHAT HAPPENED: Was it easy to match the body parts with the definitions?

APPLY: How are the parts of insects similar to animals? How are they different?

GENERALIZE TO YOUR LIFE: Why are definitions important?

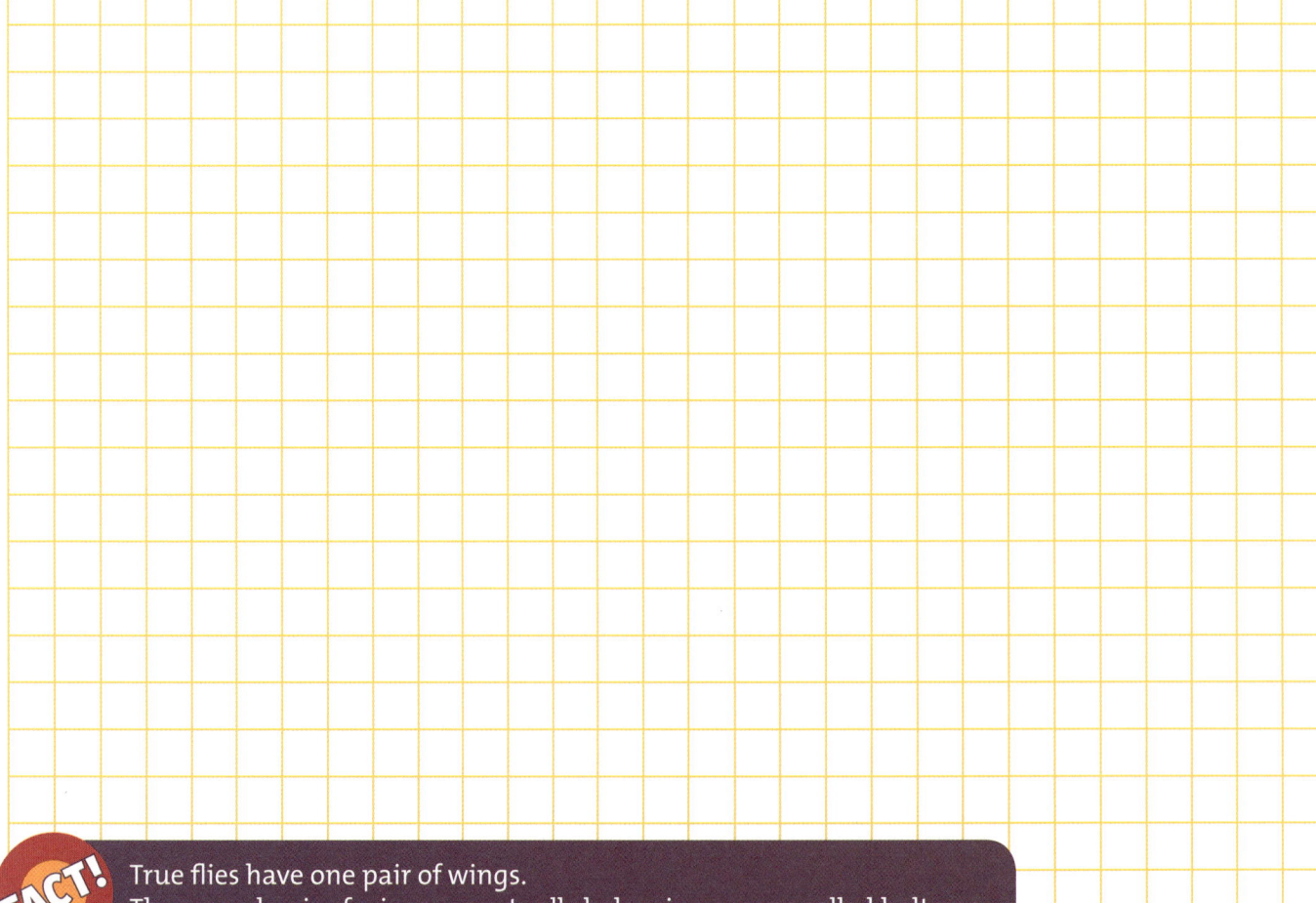

FACT! True flies have one pair of wings. The second pair of wings are actually balancing organs called halteres.

6 TEAMING WITH INSECTS

Activity 4: Big Mouth Bugs

Insects are the most successful group of animals in the world. This is because different species are adapted to many different habitats. For example, differences in mouth types allow insects to eat many different kinds of foods. In this activity you will match the mouth types with a common object that works much the same way.

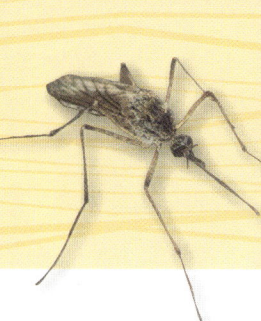

Mouths to Feed

- Draw a line from the Common Object to the Mouth Type it describes.
- Draw a line from the Mouth Type to the Insect that has that mouth type.

Common Object	Mouth Type	Insect
(sponge)	Chewing: Crushing mouth parts used to tear chunks of leaves or other types of food.	Mosquito
(syringe)	Piercing/Sucking : Long thin mouth parts used to poke into food source.	Grasshopper
(pliers)	Siphoning: Long coiled tube used to suck up liquid.	House fly
(party blower)	Sponging: Soft tissues used to mop up liquids.	Butterfly

Dig Deeper

Collect three insects and identify the mouth type each has.

Life Skills

ENTOMOLOGY SKILL: Learning insect mouth types.

SCIENCE STANDARD: Studying form and function, biological evolution.

SUCCESS INDICATOR: Ability to describe different inset mouth types.

Talk It Over

SHARE WHAT HAPPENED: Describe different insect mouth types to a friend or helper.

APPLY: How might having different mouth types help insects survive in different habitats?

GENERALIZE TO YOUR LIFE: What tools do people use to help them eat and drink different kinds of food?

TEAMING WITH INSECTS

Activity 5: FACETnating

Have you ever looked at a picture with a friend and separately described what you saw? You both saw the same thing, but you probably explained the picture differently. You learn much about your world by looking. An insect has a compound eye and sees objects differently than other animals because of the way its eye is made. In this activity you will make a viewer, look through the viewer, and get an idea of how an insect sees the world through its compound eye.

MAKE it ▶ These Eyes

Tool Kit

☐ Drinking straws
☐ Tape
☐ Red plastic sheet (clear)

When you look at a tree or person you see the whole tree or person. Insects look at a tree or person and see them divided into several thousand parts. The number of parts depends on how many facets make up each compound eye.

- Cut a handful of drinking straws in half and hold them together.
- Wrap the straws tightly with masking tape so they can stand on their own.
- You have just made a compound eye model.

USE it ▶ These Eyes

- Look at a picture through your compound eye model.
- Draw what you see.
- Find a red object or red letters and view through the red plastic.
- Look at the red object or letters through your compound eye model with the red plastic in front of it.

Life Skills

ENTOMOLOGY SKILL: Learning about insect eyes.

SCIENCE STANDARD: Insect form and function.

SUCCESS INDICATOR: Describe how an insect sees through a compound eye.

Talk It Over

SHARE WHAT HAPPENED:
- What do you see through the "insect" eye?
- How does the picture change from what your own eyes see?
- How do the red objects change when viewed through the red plastic?

APPLY: Do you think insects would be attracted to a red flower, if they can't see the color red?

GENERALIZE TO YOUR LIFE:
- Is it easier for you to describe things with words or with pictures? Why?
- If you have to describe something to someone, which helps more—a quick look or a long, careful one? Why?

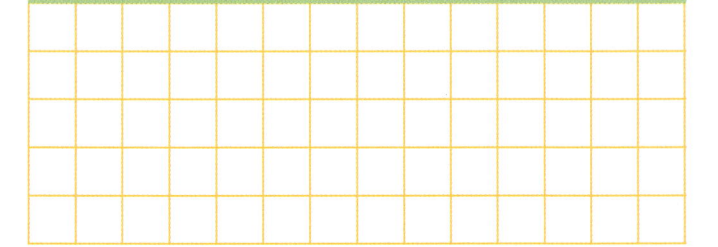

Activity 6: Insect Olympics

Human athletes compete once every four years in the Summer Olympics. Events include running, swimming, and jumping. Some athletes set new world records in their event. Insects do not have a competition in which they are tested on how fast they can fly, how far they can jump, or how fast they can swim. If insects did compete against humans in the Olympics, they might set world records! This activity helps you make comparisons between what insects and humans are able to do.

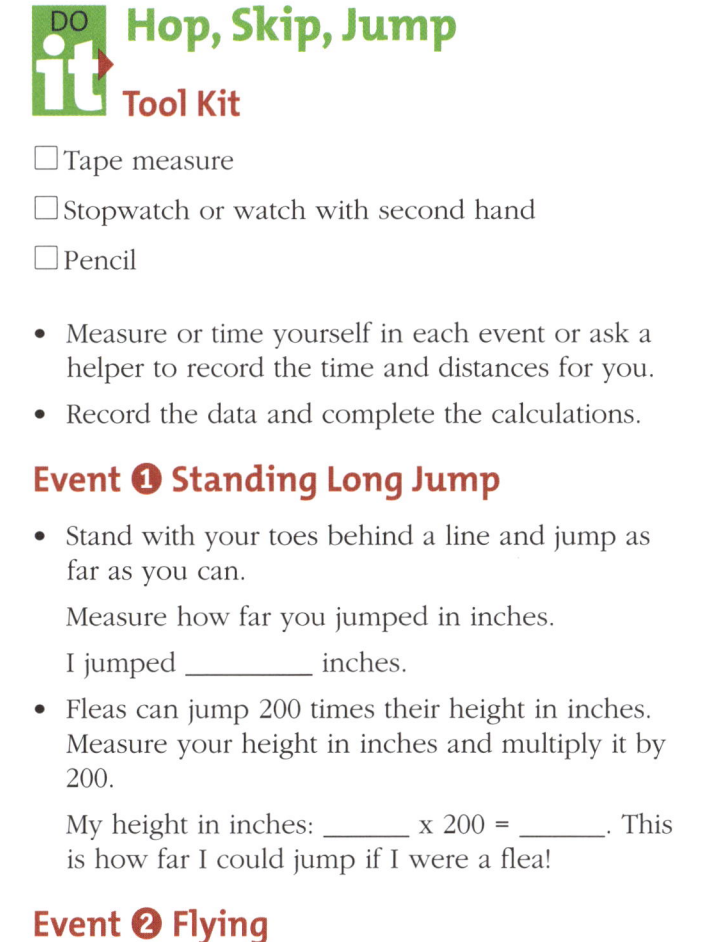

DO it Hop, Skip, Jump
Tool Kit

☐ Tape measure
☐ Stopwatch or watch with second hand
☐ Pencil

- Measure or time yourself in each event or ask a helper to record the time and distances for you.
- Record the data and complete the calculations.

Event ❶ Standing Long Jump

- Stand with your toes behind a line and jump as far as you can.

    Measure how far you jumped in inches.

    I jumped _____ inches.

- Fleas can jump 200 times their height in inches. Measure your height in inches and multiply it by 200.

    My height in inches: _____ x 200 = _____. This is how far I could jump if I were a flea!

Event ❷ Flying

- Stand with your arms out.

    Ask a helper to keep time and count how many times you can flap your arms up and down in 10 seconds.

    I can flap my arms up and down _____ times in 10 seconds.

- A biting midge can flap its wings 1,000 times in 10 seconds. Compare the number of wing flaps that you did in 10 seconds to 1,000.

Event ❸ Dash

- Measure a distance of 50 yards where it's safe to run.

    Ask your helper to time how long it takes you to run from one end and to the other.

    I can run 50 yards in _____ minutes and _____ seconds.

- A cockroach can run approximately 50 yards in 2 minutes. Compare your time in the 50-yard dash to a cockroach. Who wins the race? _____

Event ❹ Walking

- Measure the distance around the block or yard, or around the school by counting your steps as you walk.

    I took _____ steps.

    Then estimate how long your average step is in feet.

    My average step is _____ feet.

    Multiply that by the number of steps you took to determine the distance in feet.

    _____ steps x _____ feet/step = _____ feet

Now, walk that distance as quickly as you can without running. Record the time when you started and when you stopped so you know how many minutes it took you.

Activity 6: Insect Olympics

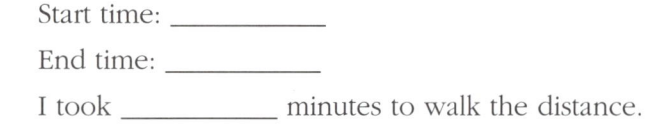

Start time: _____

End time: _____

I took _____ minutes to walk the distance.

- Ask your helper to show you how to calculate your speed (miles/hour).

To calculate your speed, divide the distance you walked by the time it took you to calculate your speed. For example, 600 steps x 2.5 feet/step = 1,500 feet. If it took you 25 minutes to do the walk, you traveled at a speed of 1,500 feet/25 minutes.

To calculate your time in miles/hour you must divide your speed (feet/min) by 5,280 (ft/mile) and multiply this by 60 (min/hour):

$$\frac{feet}{minutes} \times \frac{miles}{5,280 \ feet} \times \frac{minutes}{hour} = \ miles/hour$$

Event 5 Belly Walk

- Measure a distance of 50' in a clean hallway.

  Ask your helper to time you as you crawl on your stomach for 50' without using your hands. Record how long it took.

  Start time: _____

  End time: _____

  I took _____ minutes to crawl 50' on my stomach.

- A fly maggot (immature insect) can crawl 50' in 75 minutes. Compare your time in the belly walk to a fly maggot. Who wins the race? _____

Dig Deeper

Collect insects and their relatives such as millipedes, centipedes, and sowbugs and determine how fast they move. Make a 30.5 cm (1') long runway and release your arthropod athlete at one end. With your helper or friend, time how long it takes for the animal to travel the length of the runway. Which of the arthropods was the fastest?

You can record your data in the graph paper section below, or download a data page from the 4-H website at www.4-H.org/curriculum/.

Life Skills

ENTOMOLOGY SKILL: Learning about insect mobility.

SCIENCE STANDARD: Behavior of organisms.

SUCCESS INDICATOR: Ability to describe different insect methods of travel.

FACT! Humans have 792 muscles. Some caterpillars may have as many as 4,000 muscles.

Talk It Over

SHARE WHAT HAPPENED:
- Which event was the easiest to do? Which was the hardest?
- How did your time compare to the insect athletes?

APPLY: How does having different methods of moving help insects survive?

GENERALIZE TO YOUR LIFE: How many different methods of moving (travel) do people have?

Chapter 2: Biodiversity

Biodiversity describes the variety of animals and plants that may live together. Insects have many different (diverse) shapes, sizes, and colors. They communicate in many different ways. Insects are adapted to living in many different and unusual places. They live nearly everywhere on earth and eat many different types of foods. Some are specialists: They live only in certain places or eat only certain things. Others are generalists: They can live in many places and eat many different things.

Activity 7: Pit Stops

Have you ever gone hunting? Get ready, because in this activity you will be a hunter of very small game: insects. You won't have any trouble finding them. There are over 1 million different kinds! There are many ways to capture insects. In this activity you make and use a pitfall trap to collect insects that walk on the ground.

MAKE it ▶ Pitfall Trap

Tool Kit

- ☐ 3 plastic or foam cups
- ☐ Paper clip
- ☐ Bait (bread, fruit, lunch meat)
- ☐ 3 small pieces of board
- ☐ Small rocks
- ☐ 3 containers
- ☐ Trap Table data sheets

- Ask your helper to punch a few small holes in the bottom of your three cups with an open paper clip. This allows water to drain out.

FACT! Featherwing beetles are among the smallest known beetles. Some are less than 0.039 inches long.

DO it ▶ Pitfall Trap

Choose three different habitats to place your pitfall traps. Examples of habitats are in a lawn, under a tree or shrub, near water, in a ditch, in the woods.

- Ask for help to dig three holes and set the cups in them. Position the top of the cup even with the soil surface.
- Place your bait in the bottom of each cup. Use the same bait in each cup.
- Place a board over the top of each cup. Elevate the board above the ground by placing small rocks under it. This leaves space for insects to walk under and reach the cup.
- Empty each cup into another container each day for three days. Count the insects and other organisms collected and record them in the Trap Tables (or make your own). Add bait if necessary.
- Count the number of beetles, ants, and crickets that you collect. Ask for help if you are not sure how to identify them.

Trap Table ❶

Location: _____

Bait: _____

Date set: _____

Date examined	Total number of insects	Beetles	Ants	Crickets	Other
1.					
2.					
3.					

TEAMING WITH INSECTS

Trap Table 2

Location: _____

Bait: _____

Date set: _____

Date examined	Total number of insects	Beetles	Ants	Crickets	Other
1.					
2.					
3.					

Trap Table 3

Location: _____

Bait: _____

Date set: _____

Date examined	Total number of insects	Beetles	Ants	Crickets	Other
1.					
2.					
3.					

Dig Deeper

- Use different kinds of bait.
- Collect insects in different habitats.
- Collect insects at different times of the year.
- Learn how to use insects to make a collection at www.4-H.org/curriculum/entomology.

Life Skills

ENTOMOLOGY SKILL: Collect insects using a pitfall trap.

SCIENCE STANDARD: Learning a collecting skill.

SUCCESS INDICATOR: A collection of ground walking insects.

Talk It Over

SHARE WHAT HAPPENED:
- How many insects did you collect?
- Did you collect different types of insects in different places?

APPLY: How could a pitfall trap be used to choose the best place to put a sandbox?

GENERALIZE TO YOUR LIFE: Why is it important to collect data from different habitats?

TEAMING WITH INSECTS

Activity 8: Buzz-z-z-zing Around

People communicate with each other in many ways. Sending information, a thought, or idea can be done by writing, speaking, hand signals, facial expressions, and electronically (over the Internet). Although insects can't read or write, they have many ways to communicate with each other. They communicate to warn of danger, to tell where there is food, to attract mates, to define territories, to recognize members of the same group, and to pass other information that helps the group survive.

DO it Communication Methods

Tool Kit

☐ Pencil
☐ Small plastic comb
☐ Poster board

- Follow the instructions below to make sounds similar to insect communications.

❶ Rub your fingernail across the teeth of a comb. Describe what you hear in Your Data.

❷ Take a flashlight into a darkened room and turn the flashlight off and on. Vary the length of time for your off and on flashes to create a cycle or pattern. Describe what you see in Your Data.

❸ Hold the top and bottom of a large piece of flexible poster board. Flex the poster board back and forth faster and slower and listen to what you hear. Describe it in Your Data.

Insect Communications		Your Data
❶ Stridulation	Crickets and grasshoppers send messages by rubbing one part of the body against another. Crickets rub their two front wings together and grasshoppers rub their back pair of legs against the edge of their wing.	
❷ Light	Fireflies use light to communicate with each other.	
❸ Clicking	Male cicadas make this sound. You might hear this sound coming from cicadas sitting in trees and shrubs during July and August. Muscles in the thorax pull on a plate (similar to flexing cardboard) that flexes back and forth to make this sound.	

Insects use light, sound, and color to communicate. Examples are given below.

Communication Method	Structure Used	Function	Insect
Light	Light producing organ at end of abdomen	Attract mates	Fireflies
Sound	File on one wing rubbed against scraper on the other wing	Attract mates, protect territory	Crickets
	Abdomen tapped against wood	Attract mates	Stoneflies
	Head banged against the wall nest	Warn of danger	Soldier termites
Color	Red, black, orange, or yellow	Warn of danger	Bees, wasps, and hornets

Dig Deeper

Collect three insects and study how you think they communicate with other insects.

Life Skills

ENTOMOLOGY SKILL: Learn about insect communication.

SCIENCE STANDARD: Behavior of organisms.

SUCCESS INDICATOR: List insect communication methods.

Talk It Over

SHARE WHAT HAPPENED: How do insects communicate?

APPLY:
- How do you communicate your feelings to others?
- Why do insects need to have ways to communicate?

GENERALIZE TO YOUR LIFE: How do you communicate without words with your family or a friend?

FACT! Some female fireflies flash the pattern of another firefly species to lure and eat the male.

TEAMING WITH INSECTS

Chapter 3: Invasive Species

Animals have evolved to be in harmony with other animal species, plants, and their environment. Over time, plants and animals establish a balance. When a new species of plant or animal comes into this environment, the balance of native ecosystems can be affected. New species that harm the plants or animals in a habitat are called invasive. An invasive insect is non-native and likely to cause economic or environmental harm or harm to human health.

Activity 9: Aliens Insects

A healthy ecosystem has a balance between animals and plants competing for space, food, and water. Sometimes new species are introduced to an area that, either on purpose or by accident, upset the natural balance in the area.

Invasive Species Scramble

Unscramble the letters in the list below to complete the invasive insect names and read about the damage they cause.

Africanized _____ (eeeobhny)
More aggressive toward humans than the European honeybee, it also takes over existing hives.

Asian Long-Horned _____ (eeeblt)
Larvae destroy many species of hardwood trees by boring deep into the heartwood and robbing the tree of nutrients, eventually killing it.

Asian Tiger _____ (iooumstq)
Can transmit viruses such as Eastern equine encephalitis and West Nile virus.

Common Pine Shoot _____ (elteeb)
Larvae feed on the shoots of pine trees, causing a reduction in tree height and growth.

Emerald Ash _____ (eobrr)
Larvae tunnel under the bark and disrupt the transport of water and nutrients, eventually killing the tree.

Gypsy _____ (ohmt)
Larvae (caterpillars) defoliate (remove leaves) of many types of trees.

Light Brown Apple _____ (thom)
Larvae (caterpillars) feed on a variety of foliage and fruit crops, causing significant damage.

Mediterranean _____ (iufrt yfl – 2 words)
Attacks over 400 species of plants, including many species of citrus and vegetable crops.

Mexican _____ (rtiuf lyf – 2 words)
Larvae destroy numerous fruits of economic significance, particularly grapefruit, oranges, pear, peach, and apple, by causing them to rot.

Red Imported _____ (eifr atn – 2 words)
Can attack and cause painful stings on humans, pets, and livestock.

Russian Wheat _____ (aidhp)
Introduces a toxin as it feeds, stunting plant growth and preventing proper grain maturation in cereal crops like wheat and barley.

Silverleaf _____ (eiyfhltw)
Damages crops by feeding on them and transmitting viruses.

Sirex _____ (aoodpsww)
Feeds on several species of pine trees and introduces a fungus that kills pine trees.

Soybean Cyst _____ (aeeodmnt)
Causes stunted growth and reduced yields of soybean crops.

WORD BANK
Unscramble the words you can. Then, check them off and use the remaining words to finish the activity.

- ☐ aphid
- ☐ beetle
- ☐ beetle
- ☐ borer
- ☐ fire ant
- ☐ fruit fly
- ☐ fruit fly
- ☐ honeybee
- ☐ mosquito
- ☐ moth
- ☐ moth
- ☐ nematode
- ☐ whitefly
- ☐ woodwasp

TEAMING WITH INSECTS 17

Activity 9: Aliens Insects

Life Skills

ENTOMOLOGY SKILL: Learning about invasive species.

SCIENCE STANDARD: Behavior of organisms.

SUCCESS INDICATOR: Ability to name three invasive insect species.

Talk It Over

SHARE WHAT HAPPENED: How difficult was it to unscramble the letters to make a word?

APPLY: Do you have any of these invasive species where you live?

GENERALIZE TO YOUR LIFE: Why does the government try to keep invasive insects out of the country?

FACT! The domestic honey bee is not native to North America. English settlers brought bees to North America in 1622.

Activity 10: Establishing a Toe-Hold

All species compete to survive, but invasive species appear to have specific traits or combinations of specific traits that allow them to out-compete native species and dominate a habitat. Sometimes invasive species just have the ability to grow and reproduce more rapidly than native species. You will investigate the traits that may be responsible for the success of many invasive species.

The Competition Game

Tool Kit

- Studies show that certain traits mark a species as potentially invasive. Some common invasive species traits in non-native insects include:

    Fast growth rate.

    Reproductive strength (having many offspring and short gestation times).

    High dispersal ability (the ability to fly or spread seeds widely).

    Tolerance of a wide range of environmental conditions (generalist).

    Ability to live off of a wide range of food types (generalist).

    Association with humans.

- Think about each of these traits and complete the table below (or make your own) to explain how that trait could help a non-native insect out-compete a native insect.

Traits	How It Helps the Insect to Compete
Fast growth rate	
Reproductive strength	
High dispersal ability	
Tolerance of a wide range of environmental conditions	
Ability to live off a wide range of food types	
Association with humans	

TEAMING WITH INSECTS

Activity 10: Establishing a Toe-Hold

Life Skills

ENTOMOLOGY SKILL: Learning mechanisms of species success.

SCIENCE STANDARD: Behavior of organisms.

SUCCESS INDICATOR: Ability to list traits that assure species success.

Talk It Over

SHARE WHAT HAPPENED: Why are invasive species successful?

APPLY: Why do invasive species cost the government money?

GENERALIZE TO YOUR LIFE: What can you do to reduce the threat of invasive species?

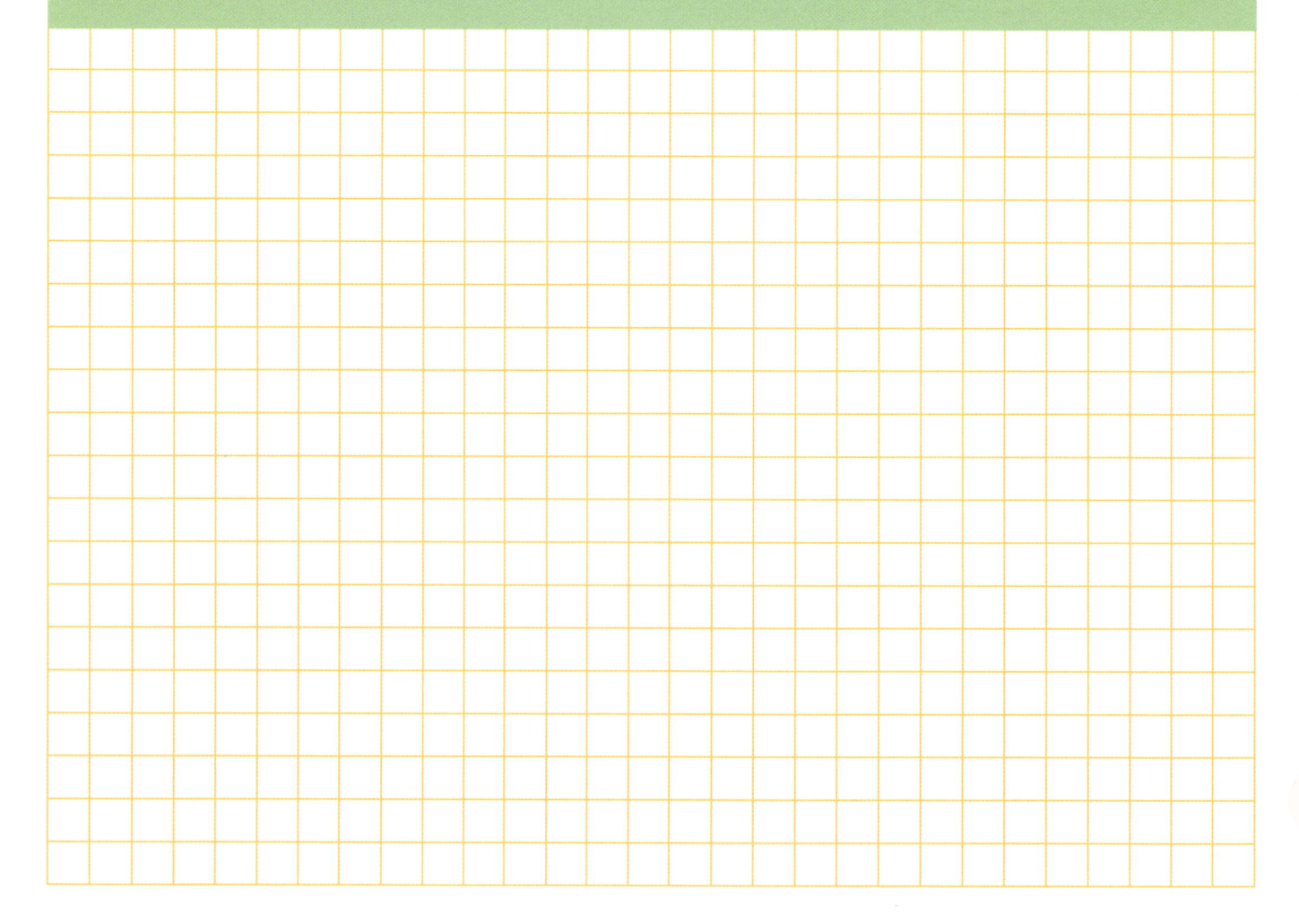

Chapter 4: Integrated Pest Management

Integrated pest management (IPM) uses all of the available methods of controlling insects that are cost effective and friendly to the environment. One of easiest ways to keep insects out of your house is to keep doors shut and screens on open windows. It is important to find out what insects are present before trying to control them. Then you can use the proper control methods for that insect.

You can help teach others how to control insects by making a poster exhibit of one of the activities in this chapter or by showing others how to do the activity in an Action Demonstration.

Activity 11: Where Are They?

Before you can begin to control insect pests, you need to find out what insects are present, where they are, and determine if the problem is big or small. In this activity you survey in and around your home for insect pests.

Finding Insects

Tool Kit

☐ Flashlight
☐ Pencil
☐ Insect Pest Inventory data sheet

- Look carefully both inside and outside your home for insect pests. Only list insect *pests*. (For example, people are happy to have a monarch butterfly on their bushes outside the house and do not want to control them, so they're not considered pests.)
- Look for signs of insect activity, such as holes in leaves, webs, anthills, tunnels, dead insects, live insects, and wood damage.
- Decide whether it's a big problem (you have to do something about it) or a small problem (you can live with it).
- Record your findings in the Insect Pest Inventory data sheet below or make your own.

Insect Pest Inventory

Location	Insects Found/Signs of Insect Activity	Problem Level	
		Big	Small

Dig Deeper

- Discuss control measures for the insect pests that you find with a helper.
- Learn how to use insects to make a collection at www.4-H.org/curriculum/entomology.

Insect Facts – Biological Control

Some insects are pests because they eat the foods we eat and live in places where we live. Some pest insects feed on humans and other animals. It is important to control insect pests without doing damage to the environment. Entomologists have developed many tools to control insects. Biological control uses other living things to control insects. Some of the major biological control agents are listed in the table.

Method of Control	Biological Control Agents Used
Predators	Lady beetles, ground beetles, mites, dragonflies, mantids, assassin bugs, robber flies, hister beetles, rove beetles, checkered beetles, blister beetles, velvet ants, spiders, lacewings, ants, damsel bugs, stink bugs, minute pirate bugs, flower flies, marsh flies, hunting wasps
Parasites	Ichneumon, braconids, chalcids, tachnid flies, humpbacked flies
Pathogens	Bacteria, fungi, viruses, protozoans

Life Skills

ENTOMOLOGY SKILL: Determining the scope of an insect pest problem.

SCIENCE STANDARD: Making observations.

SUCCESS INDICATOR: Identifies insect pests and level of problem.

Talk It Over

SHARE WHAT HAPPENED: Where did you see the most insects or signs of insect activity?

APPLY: What non-chemical control measures could help reduce insect pest populations?

GENERALIZE TO YOUR LIFE: Why is it important to take an inventory before tackling a problem?

FACT! Nearly 17% of the human population suffers from allergies to cockroaches and the house dust mite.

TEAMING WITH INSECTS

Activity 12: Ants and Uncles

Many people call any small bug an insect. Some small bugs are not insects, but they are insect relatives. In this activity, you compare insects with their non-insect relatives.

Identifying Insects and Their Relatives

Tool Kit

☐ Clear plastic cup
☐ Stiff cardboard
☐ Container with holes in the lid
☐ Insects and Relatives data sheet

- Look under rocks, leaves, boards, and other debris for insects and their relatives.
- Use a clear plastic cup and a sheet of stiff cardboard to collect ground dwellers.
- Quickly place the cup over the insect or relative.
- Carefully slide the cardboard under the cup taking care not to injure the specimen.
- Put the specimen into another container with a lid that has small holes so that the specimen can breathe.
- Collect about 10 specimens and take them home to study.
- Use the Insects and Relatives data sheet to decide which specimens are insects and which are not. A magnifying glass will help you see body part details more easily.
- Draw a picture of your specimens in the space provided.

These are NOT insects. See if you can match these Common Insect Relatives to their descriptions on page 25.

FACT! Insects and their relatives are animals (Kingdom: Animalia) that belong to the group (phylum) called Arthopoda.

24 TEAMING WITH INSECTS

Insects and Relatives

Insect Characteristics

Body divided into three regions: a head, thorax and abdomen.

One pair of antennae.

Three pairs of legs.

Some adult insects have one or two pairs of wings.

Common Insect Relatives

Spider - four pairs of legs, no antennae.

Centipede - more than five pairs of legs, body flattened, one pair of legs per body segment.

Millipede - more than five pairs of legs, body round, usually with two pairs of legs per body segment.

Sowbug - seven pairs of legs, body like an armadillo, small, gray.

Insects I Found (drawings)

Insect Relatives I found (drawings)

Activity 12: Ants and Uncles

Life Skills

ENTOMOLOGY SKILL: Insect classification, observing, collecting.

SCIENCE STANDARD: Critical thinking, social skills.

SUCCESS INDICATOR: Ability to distinguish between insects and insect relatives.

Talk It Over

SHARE WHAT HAPPENED: What insects and insect relatives did you collect?

APPLY: What are the major differences between insects and their relatives?

GENERALIZE TO YOUR LIFE: What distinguishing characteristics does a dog have?

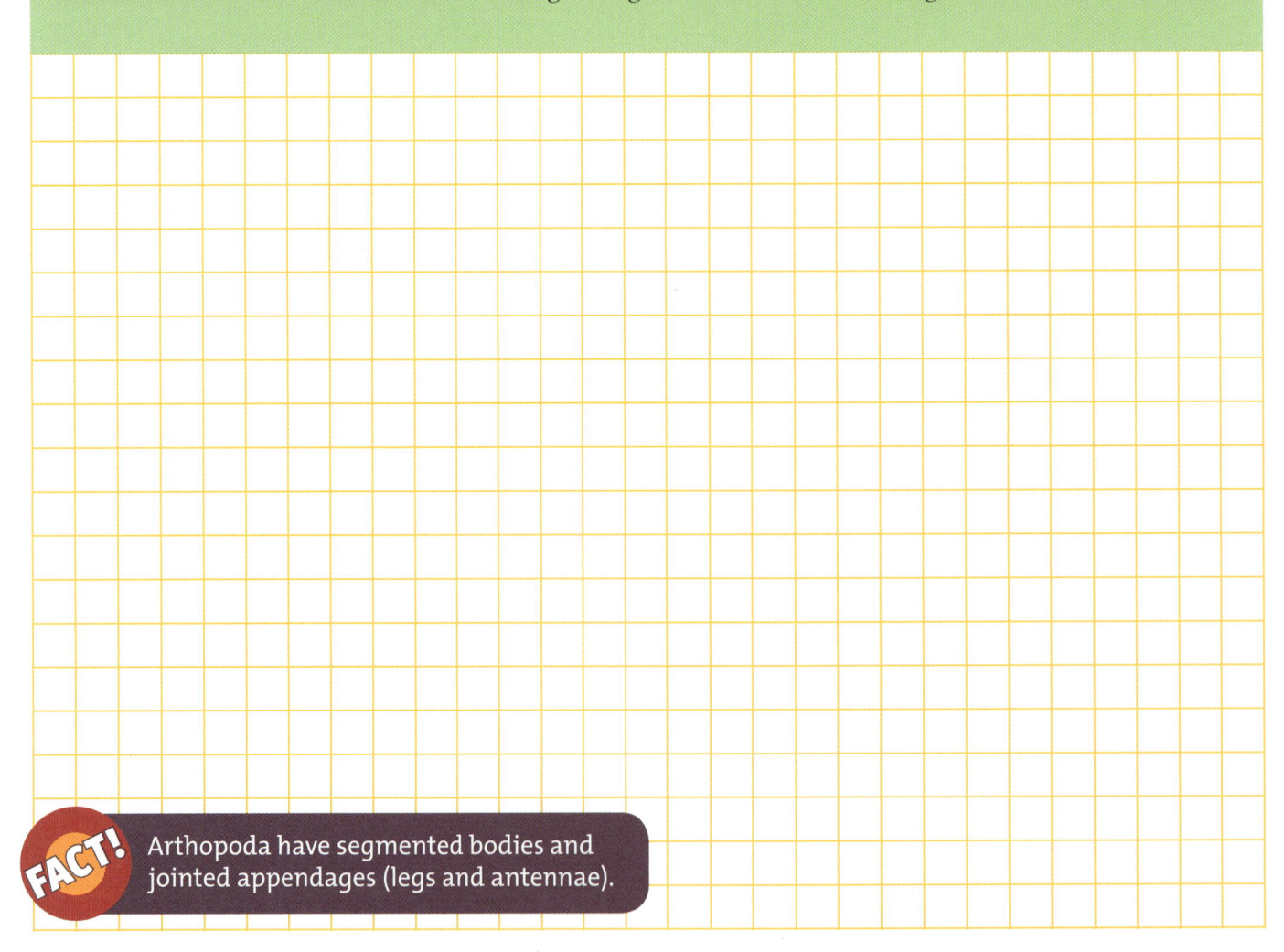

FACT! Arthopoda have segmented bodies and jointed appendages (legs and antennae).

Chapter 5: Forensic Entomology

Forensic Entomology is the science of using insects to help solve a crime. People who work in this area must learn how to make detailed observations and carefully record the information they see. Insect data can give more information to a skilled forensic entomologist than you might imagine.

Activity 13: Insect Investigation

Making careful observations is a good way to learn about insects: what they look like, what they eat, and how they behave. Good investigators record what they see right away so they don't forget.

Watch & See

- Observe insects in different habitats such as a pond, tree, prairie, field, garden, or even watch insects under a porch light.
- Pick out three different insects and observe each of them for at least five minutes.
- Carefully record your observations on the data sheet below.
- You may want to capture an insect (see Activity 12 for tips) to get a closer look. Then, you can release it or kill it quickly in the freezer to add to your collection for further study.
- Learn how to use insects to make a collection at www.4-H.org/curriculum/entomology.

	Insect ❶	Insect ❷	Insect ❸
Insect name or type:			
Date/Time:			
Weather conditions:			
Habitat:			
Description of insect:			
Behavior observed:			

Life Skills

ENTOMOLOGY SKILL: Learning about insects in their natural habitat.

SCIENCE STANDARD: Behavior of organisms.

SUCCESS INDICATOR: Recorded insect observations.

Talk It Over

SHARE WHAT HAPPENED:
- What time of day did you observe insects?
- What behavior differences did you notice among the insects?
- In what habitat did you find the most insects?

APPLY: When can taking notes be useful?

GENERALIZE TO YOUR LIFE: How do good observation skills help you communicate?

28 TEAMING WITH INSECTS

Activity 14: Chirp, Chirp

Crickets are found around buildings and in fields and meadows. Sometimes you'll find them by following their chirps. The speed (fast or slow) of their chirp depends on the temperature. You can collect crickets in the summer or do this activity anytime during the year if you buy crickets at a pet or bait store.

DO it Cricket Observations

Tool Kit

☐ 3 to 5 Crickets
☐ Clear plastic container
☐ Sand, potatoes or carrots
☐ Water bottle and cotton ball

- Collect or purchase 3 to 5 crickets. Make sure you have at least one male, since females do not chirp. (The female cricket has a long ovipositor on the end of her abdomen.) To collect your own crickets, do it in late summer or early fall when adult crickets appear. Use an insect net to gently capture crickets.
- Put a shallow layer of sand in the bottom of a small, clear container. Punch small holes in the lid of the container so the crickets can breathe.
- Place small pieces of potatoes or carrots in the container for the crickets to eat.
- Place a small bottle filled with water and a cotton ball plug so the crickets have water. Place the crickets in the container and secure the lid.
- Watch and listen to the crickets for five minutes, three times a day, for three days. Include day and night observations. Record what you see and hear.

Day	Time	Sounds (# chirps/minute)	Observations
1			
1			
1			
2			
2			
2			
3			
3			
3			

TEAMING WITH INSECTS

Activity 14: Chirp, Chirp

Dig Deeper

- Repeat this experiment at a different temperature. Avoid extremely hot or cold temperatures that can kill the crickets.
- Watch your crickets in total darkness using a small flashlight or lamp with a low wattage red light bulb. Insects cannot see red and the crickets will think it is dark. Observe any differences in cricket behavior in the light and in the dark.
- Learn how to use insects to make a collection at www.4-H.org/curriculum/entomology.
- You can record your data in the graph paper section below, or download a data page from the 4-H website at www.4-H.org/curriculum/.

Life Skills

ENTOMOLOGY SKILL: Observe cricket behavior and communication.

SCIENCE STANDARD: Communication.

SUCCESS INDICATOR: Describe how insect chirps are affected by temperature.

Talk It Over

SHARE WHAT HAPPENED: What cricket behavior did you see?

APPLY: Can you tell the temperature by counting the number of chirps from a cricket?

GENERALIZE TO YOUR LIFE: Why do you think male crickets communicate differently depending on the weather?

FACT! There are 900 species of crickets.

Activity 15: Sherlock Bug

Insects are everywhere, but we might not know it because we cannot always see or hear them. A careful look around can reveal quite a few outdoor insect hiding spots. In this activity, you are a bug detective looking for insects and clues that lead to their hiding places.

Signs of Insects

- Plants are great places to look for insect activity. Holes in leaves, petals, or on other plant parts may show the presence of chewing insects.
- Examine the stem or leaves of plants for galls. Galls are large swellings or wart-like growths that are commonly found on oak and maple leaves and on goldenrod stems.
- Look for long twisting tunnels on plant leaves. These tunnels are made by leaf mining insect larvae that feed between the top and bottom layers of the leaf.
- Find an old log and peel away the bark. On the surface of the wood you should see some tunnels made by beetle larvae tunneling through the wood. Replace the bark when you are finished with your observations.

Insect Observations

- Search outside for four different clues that reveal the presence or activity of insects.
- Record the clue you found for each insect and which insect you think left it in the chart below (or make your own).

Case 1	Case 3
The insect clue:	The insect clue:
I think the insect could be:	I think the insect could be:
Case 2	Case 4
The insect clue:	The insect clue:
I think the insect could be:	I think the insect could be:

Activity 15: Sherlock Bug

Dig Deeper

Go outside on a summer evening. Close your eyes and listen. What insect sounds do you hear? Record all the sounds you hear during three nights.

- Repeat this activity under different weather conditions.
- Repeat this activity at different times of the night.
- You can record your data in the graph paper section below, or download a data page from the 4-H website at www.4-H.org/curriculum/.

Life Skills

ENTOMOLOGY SKILL: Learning to infer what insects are present without seeing the insects.

SCIENCE STANDARD:
- Abilities necessary to do scientific inquiry.
- Understanding about scientific inquiry.

SUCCESS INDICATOR: Identification of four insect clues.

Talk It Over

SHARE WHAT HAPPENED: What insect clues did you find?

APPLY: What attracts insects to different places?

GENERALIZE TO YOUR LIFE: What clues do birds and mammals leave?

FACT! Tannins, extracted from insect galls, were used to tan hides and to make permanent inks.

Activity 16: I Eat Insects

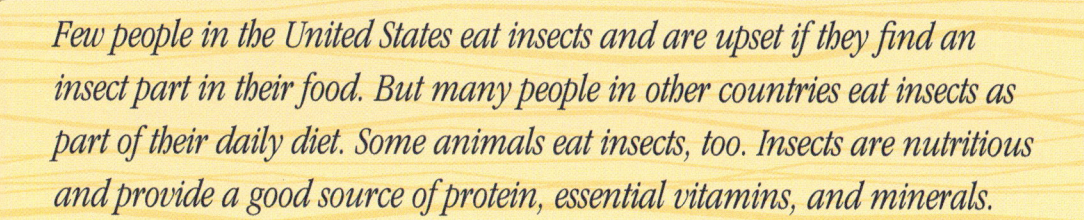

Few people in the United States eat insects and are upset if they find an insect part in their food. But many people in other countries eat insects as part of their daily diet. Some animals eat insects, too. Insects are nutritious and provide a good source of protein, essential vitamins, and minerals.

Insect Eaters

Read the clues and list the insect-eating animal described.

Clue ❶ I wear a fur coat, have claws, and sleep all winter. I love honey and eat insects. What am I?

_____

Clue ❷ I fly at night and find insects to eat by using echoes. Sometimes I live in a belfry. What am I?

_____

Clue ❸ I have scales and fins and love to swim. I eat worms and insects. What am I?

_____

Clue ❹ I hop and live in ponds. What am I?

_____

Clue ❺ You have two legs and I have 100. So which foot do I move first? What am I?

_____

Clue ❻ I have feathers but can't fly very far. I give you eggs for breakfast. What am I?

_____

Clue ❼ I have a long snout and a long tongue to catch ants. What am I?

_____

Clue ❽ I am a plant. Some people think I can play baseball because of my name. Insects fall inside and are caught in a liquid at the bottom. What am I?

_____

Clue ❾ I sat down beside her and Little Miss Muffet ran away. What am I?

_____

WORD BANK
Check off each insect name as you enter it in one of the blanks.
- ☐ anteater
- ☐ chicken
- ☐ fish
- ☐ bat
- ☐ spider
- ☐ millipede
- ☐ frog
- ☐ pitcher plant
- ☐ bear

Activity 16: I Eat Insects

Insect Facts – Insects People Eat

People in other countries often eat insects as a good source of protein. Some of the insects eaten in selected countries are given in the table below.

Country	Insects
Mexico	Aquatic bug eggs, caterpillars of giant skipper butterflies, ant larvae
Africa	Grasshoppers, termites, caterpillars of giant silkworm moths, weevil larvae and adults, crickets, green shield bugs, black ants, honeypot ants, cicadas, palm rhinoceros beetles
Southeast Asia	Giant water bugs, brown crickets, silkworm pupae, wasp larvae, caddisfly larvae, cicadas, grasshoppers, dragonflies
South America	Leafcutter ants, beetle grubs, palm weevils

Life Skills

ENTOMOLOGY SKILL: Reviewing animals that eat insects.

SCIENCE STANDARD: Behavior of organisms

SUCCESS INDICATOR: List animals that eat insects and insects that some people eat.

Talk It Over

SHARE WHAT HAPPENED: How many animals could you name after reading the clue?

APPLY: Why do some animals eat insects?

GENERALIZE TO YOUR LIFE: Why do different people eat different things?

Resources

Websites for the following resources may be accessed at www.4-H.org/curriculum/entomology.

Publications

Arthropod Collection and Identification: Laboratory and Field Techniques. T. J. Gibb and C. Y. Oseto. 2006. Academic Press, New York.

Borror and DeLong's Introduction to the Study of Insects. 2005. C. A. Triplehhorn and N. F. Johnson. 7th ed. Thomson Brooks/Cole, California.

Bug Hunter. 2005. David Burnie. Dorling Kindersley, New York.

Bug Lifecycle Fun Book. 2005. Insect Lore.

Butterflies of North America. Brock, J. P. & K. Kaufman (2003), Kaufman Focus Guides, Houghton Mifflin Co., New York.

Familiar Butterflies of North America. Walton, R. K. (1990), National Audubon Society Pocket Guide.

Field Guide to Insects America North of Mexico. 1970. D. J. Borror and R. E. White. Houghton Mifflin Company, New York.

How to Make an Awesome Insect Collection. 2009. T. J. Gibb and C. Y. Oseto. Purdue Extension No. ID-401. Purdue University, West Lafayette, Indiana. Available online at www.4-H.org/curriculum/entomology.

Insect Almanac: A Year-Round Activity Guide. 1992. M. Russo. Sterling Publishing Co., New York.

Insect Life: A Field Entomology Manual for the Amateur Naturalist. 1985. R. H. Arnett and R. L. Jacques Jr. Prentice-Hall, New Jersey.

Insect. 2004. L. Mound. Dorling Kindersley, New York.

Insects and Spiders. 1998. J. VanCleave. John Wiley & Sons, New York.

Looking at Insects. 1992. D. Suzuki. John Wiley & Sons, New York.

Mosquito Bite. 2005. A. Siy and D. Kunkel. Charlesbridge, Massachusettes.

National Audubon Society Field Guide to North American Insects & Spiders. 1980. L. Milne and M. Milne. Alfred A. Knopf, New York.

National Wildlife Federation Field Guide to Insects and Spiders & Related Species of North America. 2007. A. V. Evans. Sterling Publishing Co., New York.

Online: Insect. 2007. I. Graham. Dorling Kindersley, New York.

Pet Bugs: A Kid's Guide to Catching & Keeping Touchable Insects. 1994. S. Kineidel, John Wiley & Son, New York.

The Practical Entomologist. 1992. R. Imes. Simon & Schuster Inc., New York.

Professional Societies and Organizations

Amateur Entomologists' Society

Entomological Society of America

National Pest Management Association, Inc.

National Invasive Species Information Center

Extension Websites

Websites for the following resources may be accessed at www.4-H.org/curriculum/entomology:

The Beehive

Purdue University, Department of Entomology

eXtension, Pest Management

How to Make an Awesome Insect Collection

TEAMING WITH INSECTS

Glossary

Abdomen Often the largest region, it usually has nine or 10 segments or rings.

Adaptation A change that occurs in an animal's behavior or body that allows it to survive and reproduce in new conditions.

Antennae Moveable structures found in pairs on the head of insects used to hear, taste, and smell.

Arthropod A group in which animals have jointed legs, a hard outer covering, a body divided into rings, and both left and right sides the same.

Biodiversity Variations of animals and plants in a habitat. It is often used to measure the health of biological systems.

Biocontrol Control of pests using other organisms.

Bug A term used to refer to insects in general. The term bug should be used only for members of the insect order Hemiptera, the true bugs.

Compound eye An eye comprised of many smaller eyes, each with its own lens. Compound eyes are found in most insects.

Diversity Differences or variety.

Ecosystem Includes all the plants, animals and the physical, non-living things in an area. A healthy ecosystem has a balance between the living things.

Entomologist A scientist who studies insects.

Entomology The scientific study of insects.

Evolution The gradual change over many generations in plant and animal species as they adapt to new conditions or new environments.

Facet Outside surface of one lens of compound eye.

Forensic entomology The study of insect and arthropod biology to assist in police investigations.

Forewing The first pair of wings; can be thin or thick, transparent or colored, scaly or smooth.

Gall A growth in plant tissue often caused by insects or mites feeding or laying their eggs.

Habitat The natural home of an animal or plant.

Halteres Modified wings that form drumstick-shaped balancing organs.

Head The first of three insect body regions. The head contains mouthparts, eyes, and antennae.

Integrated pest management Also known as IPM. Management of insects using a variety of control measures based on economics and causing the least harm to the environment.

Insect An arthropod with three body divisions, three pairs of legs, and with one or two pairs of wings.

Invasive species An insect, plant, or animal that is non-native and likely to cause economic or environmental harm or harm to human health when introduced.

Larva A young animal that looks completely different from its parents. Some insect larvae (the plural form) change into adults by complete metamorphosis. A beetle larva is sometimes called a grub.

Metamorphosis Change in shape or form during the immature to adult stages (caterpillar to butterfly).

Millipede Not an insect (insect relative). Sometimes called "thousand-leggers" because of the many legs, Millipedes are arthropods and closely related to centipedes. Each body segment has two pairs of legs. They feed on decaying plant materials.

Ovipositor An egg-laying structure located at the end of the female's abdomen.

Species A group of similar living organisms that mates and produces young.